Game Theory for Business:

A Simple Introduction

Also by K.H. Erickson

Simple Introductions

Choice Theory
Financial Economics
Game Theory
Game Theory for Business
Investment Appraisal
Microeconomics

Game Theory for Business:

A Simple Introduction

K.H. Erickson

© 2013 K.H. Erickson

All rights reserved.

No part of this publication may be reproduced, stored in or introduced into a retrieval system, or transmitted in any form or by any means, including electronic, mechanical, photocopying, recording or otherwise, without the prior permission of the author.

Contents

Introduction	6
Business Goals	9
The Business's Dilemma	15
Adverse Selection and Moral Hazard	26
Repeated Gains	40
The Business of Show	49
Business Competition	59
Tacit Collusion	69
Entry Deterrence	80
Strategic Behaviour	87
Firm Survival	92
Niche Businesses	99

Introduction

Game theory is a useful tool to model interactions and gain insight into what may appear to be unpredictable or unsatisfactory outcomes. The theory sees every interaction as a game, where the players involved aim to select the best strategy available to achieve the highest possible payoff on offer. In practice this simplifies to a choice of just two strategies for a player, as they either act a certain way or they don't. But a game never contains only one participant, and those involved must take account of the presence of those in a similar position or risk being played. A game can force a poor payoff on a player if they only look at the game as a whole and ignore the individual incentives facing others, although trying to outmanoeuvre a rival can also backfire as both players rush to the bottom and end up worse than they started.

The characteristics of game theory make it especially relevant to the business world, and in a more general sense to all those trying to buy or sell something. A business manager may wonder why profits and sales are lower than expected, despite selling a desirable product and using a well thought out pricing strategy to appeal to customers and outmanoeuvre competitors. And a buyer may feel unsatisfied with their interactions with a business, despite paying large sums of money and being a likeable customer. Game theory can offer insights into all of these scenarios and more, to look in depth at the incentives facing the other side and suggest strategies to take advantage of them.

This book offers a simple and accessible introduction to the central ideas and principles of game theory for business. The games are illuminated with over 35 diagrams used throughout to show business theory and games in action, as both forward and backward induction are utilised to predict a business's payoffs. Dominance and the Nash equilibrium are explained and illustrated to show how business strategies may be formulated in practice.

First the main wealth seeking priorities for a business are put forward and justified, to lay the foundations for the prisoners' dilemma game a business or seller of any product must face. Several possible solutions to the game are explained, as a business may operate a bait and switch strategy as a short-run tactic, or build trust with a customer in an attempt to create long-run gains. The case of show business is also investigated, where a business may be able to escape the prisoners' dilemma game other firms struggle against and create a game built around their own best interests.

Rival businesses are the focus for the remainder of the analysis, looking into a firm's pricing incentives and at the prospects for collusion to achieve higher profits, before turning to the strategies available to repel new competitors planning to enter a firm's industry. A business's situation is compared to similar scenarios in evolutionary biology and lessons are drawn for what this can teach sellers competing for sales, and some specific strategies for a business trying to stay ahead of the competition are explained.

Business Goals

The goal of any business is first and foremost to maximize its wealth. This is true across all cultures worldwide, in every business sector, and holds irrespective of the current financial status of a firm, no matter if it's close to bankruptcy or at the very top of its field. A large corporation is focused upon creating wealth for the shareholders who own it, while a small one person business based around self-employment is also primarily concerned with wealth maximization. Even if a business's interests are focussed in the charity sector, the most important goal has to be to increase its own wealth, as without it the economic future and survival of the firm is at risk, and they may not be around to work on their mission to help others.

Just as a business intends to gain wealth to assure its own survival as a prerequisite for going after its goals, each and every individual is trying to achieve the same thing. In a world where money doesn't grow on trees one person can only gain it by taking it from someone else, and this creates a stalemate situation where a business aims to take money from individuals, while they all target the opposite and plan to hold onto their cash. But this doesn't have to result in an armed fight to the death, where the winner claims the money needed to assure their own long-run survival, and the loser lies motionless in a heap. Instead it's possible for a fair exchange to take place, where a sum of money is swapped for something of value, which is worth more to the potential buyer than the money that would be given up. And of course it follows that the business will only part with the product or service if it believes that the money on offer is more valuable.

As the two different sides in any business transaction have opposing goals, every interaction between a business and their customers is destined to turn into a battle over value. The firm seeks to raise the value of their product to claim more money, while the customer wants to raise the value of their money and get a bargain. And the result of the interaction between a buyer and a seller may be a stalemate, where the seller refuses to lower their prices, and the buyer refuses to increase the amount they're willing to pay.

But if neither side in a businesses transaction ever budged then no trades or transactions would ever take place in an economy, and every individual would only have the products that they themselves currently possess, and nothing more. There would be no reliable way to gain a house, a car, food, or any other essential resource, as no-one would be willing to make an exchange in return for them, and civilization as we know it would cease to exist. Society would return to the law of the jungle, where the only way to survive is to use brutal violence against others and pillage their resources, and life would be short and worthless.

The diagram represents a situation where it's every man for himself, and where the only way to get an essential resource is to steal from others. It shows the possible outcomes of an interaction between two different people, person one who has the rows across and the first payoff of each pair, and person two who has the columns down and the second payoff of each pair. Each person has a choice of two different strategies, A and B. If both select the same strategy then person one gains a positive payoff and person two suffers a negative payoff, as person one is more suited to direct competition. But if both people select different strategies this is reversed as person two gets the better of things, being more suited to indirect competition.

A zero-sum world

	Person 2 strategy A	Person 2 strategy B
Person 1 strategy A	-1 1	2 -2
Person 1 strategy B	3 -3	-4 4

But no matter which of the four outcomes here occurs the overall payoff for society, found by combining the two people's payoffs, is the same:

Strategy 1A and Strategy 2A = 1 + (-1) = 0
Strategy 1A and Strategy 2B = (-2) + 2 = 0
Strategy 1B and Strategy 2A = (-3) + 3 = 0
Strategy 1B and Strategy 2B = 4 + (-4) = 0

All of the possible outcomes here are zero-sum as the payoffs total to zero, and this represents a situation where there is no growth and no development in a country. To avoid this unwanted situation and for the greater good of a well-functioning society an alternative is required, where mutually beneficial interactions can occur with the payoffs between both sides summing to more than zero.

If direct violence and intimidation is considered off-limits in a civilized society then this force will have to be indirect, through manipulation. The business seller can use any trick to convince others that its products are high value, and far more valuable than a sum of money, while the customer and potential buyer can act like the product has no value to them, although they may be interested if the price was low enough. The result is that one side may 'win' and trick the other party, or alternatively neither or even both might win and get what they want, depending on how the buyer and seller value the product for sale and the sum of money that's being offered for it.

A situation where an individual attempts to manipulate a business each and every time they require a different good or service is going to be inefficient, and if the potential buyer is unsuccessful in getting the business to lower the price then time has been wasted and a beneficial interaction hasn't occurred. The same holds if a business targets potential customers one at a time, with the intention of raising the price they charge, and this may also waste time and effort with no return. Where buyers and sellers interact one to one an interaction will be based around haggling and bargaining, and this is especially common outside of the wealthy West. But repeated bargaining will inevitably involve a great deal of failed attempts and wasted resources as a result, and it's unlikely to be a coincidence that haggling and bargaining are most common in the economically poorer countries.

The best scenario to see beneficial transactions proceed would involve widespread manipulation attempts occurring at the same time. This way there's a greater chance that at least some of the attempts would be successful, to create the transactions a functioning society requires. There's no way for a customer to manipulate multiple businesses at the same time as firms will be located in different space, and even if they weren't they aren't going to collectively give an audience to random individual customers. However, it is possible for a business to manipulate large groups of people at the same time through the use of marketing or advertising. If multiple businesses were to advertise and market their products well then significant numbers of transactions and trades should result, to allocate resources across communities and maintain a functioning civilized society for the greater good of all. The goal of every business is therefore to do whatever they can to manipulate large numbers of customers into buying their products, and to win the prized sales and money they need to survive. But to have a chance at the prize a business must first play the game.

The Business's Dilemma

A business's goal is to raise the price of the sale while a buyer's goal is to reduce it and this tug of war becomes a game, with the winner being the player that achieves their goal and the loser the player that suffers the opposite. It's also possible for both players to achieve a small victory, with a seller getting the minimum amount he was prepared to accept for his product, and the buyer gaining a good that is valued slightly more than the sum of money used to acquire it. The final possible outcome of the game is for neither player to win or lose, as the business transaction doesn't take place and both the business and the potential customer walk away, ending up exactly as they started.

A game's four potential outcomes can be given numerical payoff values, to represent the relative effects they may have on a player. If neither of the two players budges and the interaction doesn't proceed then both the business and the customer have payoffs of 0. Alternatively, both sides may claim a modest victory with an acceptable price for buyer and seller, and this may give payoffs of 5 each.

If the business uses a good marketing strategy to manipulate a customer to pay far higher than he intended then the business gains a sum of money far greater than their minimum selling price. This may give a large payoff gain of 10 for the selling firm, and a payoff loss of -5 for the tricked and annoyed buyer who parted with far more money than was required and now feels foolish, although his loss is not equal to the seller's gain as he does get a product of some value just not what he'd hoped.

The interaction may also take the opposite path instead, with the buyer the player who manages to manipulate the business seller into making the sale at a discount, perhaps convincing the business that their products are inferior to what they're claiming. If this happens then the buyer pays far less for the product than he was willing to for a payoff of 10, and it's the business who suffers the negative -5 payoff as they gain a significantly lower sum of money than their minimum selling price and less than they could have made. But again the firm's payoff loss is not equal to the buyer's gain as they do gain a sale and some profits.

The specific number values given to an outcome are not important here, and what matters is their relative value and ranking. A player's best outcome is a large win as the other player loses, second best is to share a modest victory, the second worst result is for a stalemate with no transaction, and the worst outcome is to suffer a payoff loss as the other player wins.

All of the above information can be put into a clearer diagram form, in what is called a game matrix. This shows all of the important information relating to a game, including the players involved, the strategies available to them, and the corresponding payoffs linked to each possible outcome. The business seller's options and corresponding payoffs, depending on the buyer's choice, are in the rows across, with the buyer's potential actions and the resulting payoffs, linked to the business's behaviour, in the columns down. The matrix has the same format as the zero-sum game earlier and shows the separate strategies that a buyer or seller may follow here, linking the four possible outcomes to a specific combination of behaviours by a business and customer.

	Buyer compromises	Buyer raises demands
Seller compromises	5 5	10 -5
Seller raises demands	-5 10	0 0

A scenario where there's no trade or where one player successfully manipulates the other will only come about because a player refuses to compromise and instead makes unreasonable demands. This would involve the seller insisting on a price far higher than the minimum they would be willing to accept, or the buyer pushing for a price far lower than what he's willing to pay. If the other player also makes unreasonable demands in a similar way then there will be no trade (payoffs of 0 each). But if they were to compromise and go with what the demanding player wants, not noticing the fact that the requests are ridiculous and totally at odds with their own needs, then they will suffer a negative outcome (-5 payoff), while the other player's bad behaviour is rewarded (10 payoff). And if both players compromise on their excessive demands then they share modest gains (5 payoffs each).

To predict the result of this game the matrix above can be separated into smaller parts. This shows how one player can be expected to react given a certain strategy from the other, assuming players seek a best response and the highest possible payoff. The following diagram takes a part of the larger game matrix above to show how a business seller will respond to a customer's behaviour.

	Buyer compromises
Seller compromises	5
Seller raises demands	10

If the buyer was to adopt a compromise strategy during the interaction, giving up on their desire to lower the price and instead pushing for a deal, then the seller would be better off raising its demands than copying the attempt to cooperate. This will give a payoff of 10 instead of 5 here, and although the relative payoff values might be different in real life, it's reasonable to assume that more can be demanded if the other person is committed to a compromise strategy.

	Buyer raises demands
Seller compromises	-5
Seller raises demands	0

In the alternate scenario where the buyer raises their demands of the seller, pushing relentlessly for a lower price, the best response for the business is to raise its own demands and not try to compromise. This would see the end of the interaction as the two parties go their own way, giving payoffs of 0 each as neither money nor a product changes hands. But the firm would avoid the loss that is associated with compromising with a demanding customer, a negative payoff of -5 here.

Irrespective of what a customer does, whether they raise their demands or attempt to compromise, it's best for a business to raise their own demands and push for a higher price. If they're successful they achieve a superior payoff with more money, and if they're not successful and the customer leaves then they end up no worse than they started, and live to play the game another day. There is no chance of a negative payoff this way, while the compromise policy puts a business at risk of a loss.

The payoffs for the customer here are identical to those of the business and therefore their strategy doesn't need to be examined separately. Even if the buyer's payoff values were slightly different to the business seller's the relative payoffs would remain the same, and it will be best for the customer to raise their demands irrespective of the business's behaviour, pushing the buyer to ignore everything the business says or does that doesn't go along with their wishes. Such a strategy will see the buyer pay a lower price and save money in the best case scenario, or leave without incurring a loss in the worst case scenario.

Putting it all together, it appears that both the business and their customers can be expected to raise their demands in any interaction, and resist attempts to compromise. The compromise strategy is strictly dominated by the strategy to raise demands, and given a choice by the other player a decision to compromise can not only never offer a higher payoff than raising demands, but it will also always offer a lower payoff. This gives a Nash equilibrium expected outcome where both players raise their demands and neither gets what they want from the other, and the sale does not occur. This result is underlined in the game matrix below, with the payoffs of 0, 0 showing that neither party gains anything from the interaction.

	Buyer compromises	Buyer raises demands
Seller compromises	5 5	10 -5
Seller raises demands	-5 10	<u>0</u> <u>0</u>

The natural result of a seller-buyer interaction sees no hope for a transaction. However, if players had been able to achieve a mutual compromise they would have received payoffs of 5, 5 with a gain for each. This is therefore a prisoners' dilemma game, where the two players would both be better off if they cooperated but individual incentives prevent it. The lack of trust or a binding agreement between the two players makes cooperation impossible.

In the previous section it was explained that a society needs interactions between businesses and customers, in order to generate wealth and maintain a civilized and functioning economy. Businesses in particular are expected to take part in large numbers of transactions to achieve this end, and they're in the unique position of being able to target large numbers of potential customers at once to potentially bring about multiple transactions with little time and effort. But with customers well aware of a business's incentives to raise their product's price to more than its worth, and their own incentives to not budge on their demand to lower the price, it appears that a business will struggle to achieve any sales at all. This is the business's dilemma, and a firm's individual incentives appear to work against its best interests to achieve a higher payoff.

Yet despite what this game would suggest in real life businesses are able to sell their products to demanding customers and still earn good money. It appears that something is missing from the analysis. It's useful to remember that the game shown above is a prisoners' dilemma game, based on a model where two gang member prisoners are questioned by police over their crimes. If the two gang members keep quiet and remain loyal to each other there's only evidence to charge them with a lesser crime with short sentences, but their individual incentives are to chase a better deal with the police and give evidence against the other man to save their own skin. The expected result of the game is for each player to follow their incentives to throw the other under the bus, just as here, giving the police enough evidence against each of them to see both serve the maximum sentences in jail. But the actual result may be different.

In real life some criminals do collectively keep quiet and enjoy superior shared payoffs, with seasoned convicts in particular often aware of the bigger picture. They value the prolonged long-run payoffs they'd receive in return for gang loyalty, and value this more than the payoffs linked with reduced jail time. It follows that the way for a business to bring about a successful transaction with a customer is to look to the long-run.

The game matrix above is a one-off static representation of a seller-buyer interaction and the repeated dynamic game will be different. In practice the outcome where one player 'wins' and screws over the other can only happen once, and after that a deceived buyer would not trust the seller anymore, while a manipulated seller would not be willing to do further business with the problematic customer. As a result the highest possible payoff of 10 here is only available once before the loser suffering the -5 payoff walks away, but the mutual compromise outcome could be repeated countless times, each offering a 5 payoff for both players as noted. With enough repeated compromise transactions the returns to both customer and business would significantly outweigh those from a one-off interaction. Therefore the choice for players is to either target a big win over the other player for a high one-off payoff, before searching out for another buyer or seller to repeat the process, or target a lower payoff through compromise and not have to constantly search for new sources of trade.

It will be difficult for a buyer to repeatedly find suitable new businesses to interact with, and then manipulate into agreeing to a lower price than they had intended. On the other hand it will be far easier for a business to find ever more customers, and the typical form of interaction is for the firm to advertise its products or brand and wait for customers to come to them. This gives a business an advantage, and while it may be willing to screw over buyers with a high price and count on always being able to attract new suckers, a customer doesn't want to have to spend time and effort searching out and manipulating new businesses. Unlike the firm a buyer's life is not set up to make this their focus, as they will have a full-time job to occupy their time and a separate life to lead.

Due to the time and effort required to constantly find new businesses to deal with a customer is likely to prefer to build a relationship with a reliable business, and select the mutual compromise option. This will offer a payoff that may be lower individually, but which can be gained repeatedly to offer reliable long-run savings and generate value and wealth. But customers will only do this if they believe they aren't going to be exploited, and think that the business will also cooperate and compromise. A business's success in playing the game and achieving the best payoff for itself will therefore depend entirely upon convincing customers that they intend to compromise, and that they value the repeated modest gains associated with a long-term business relationship over higher one-off gains.

Adverse Selection and Moral Hazard

As consumers are the lifeblood of a business a successful firm will need to see things from their point of view. Consumers have two main fears that may hold them back from a business transaction, and these fears are represented by the ideas of adverse selection and moral hazard.

Adverse selection occurs when consumers possess imperfect information about a business or its products, and this lack of full information means that a buyer can't be sure if the firm is a high quality business with high quality products and service, or a low quality business with low quality products and service. A buyer would pay a high price for a high quality business transaction, but only a low price for a low quality business transaction, yet with incomplete information about sellers and their products a buyer can't be sure which is which and must treat them as one group.

A high price a buyer will pay for high quality is averaged out with the low price they'd offer for low quality, for a medium price for the overall medium average quality in the market (some high quality products and some low quality gives medium quality overall). But a medium price is less than a high quality seller is worth and this sees them withdraw from the market, while the medium price is above what a low quality seller is worth and this attracts them to the market in large number. Soon enough only low quality sellers are left in the market and a buyer can't find good quality businesses or products, causing the adverse selection problem. Even if the issue of price is ignored a low quality seller will naturally find it harder to make a sale than a high quality seller, and this will leave them more desperate and more likely to take part in the market than a high quality seller, ensuring that they are what a buyer is more likely to find.

Moral hazard is similar to adverse selection but occurs after the transaction has started instead of before. A firm may intend to offer high quality products or service, but once the buyer has handed over the money the firm may realize that they can no longer face negative consequences by lowering the quality of their business offering. With the buyer the only one to suffer from it they may go ahead and reduce effort and investment for a worse service or products than were promised.

If a firm is a high quality seller then it may be able to find a way to show indicators of this, and tie itself into a binding contract, to help a customer overcome their fears of adverse selection and moral hazard to see a business transaction occur. But a low quality seller won't be able to do this, and they may be unwilling if an immoral firm which wants to keep their options open to act dishonesty in the future, or simply unable if a new up and coming business that is still learning their trade. But a consumer is unlikely to proceed with a business transaction without some reassurances, and a business that can't offer the real thing will need to create some believable bait to mimic the appearance of cooperation.

A business will basically have to lie, and create deceptive bait out of nothing as part of a bait and switch strategy. The firm may justify this behaviour with the idea that their country needs new businesses to survive to boost the economy, and a business manager may insist that the lies and deception are only a temporary phase, until they have generated enough sales and wealth to build up the firm and give customers the value they desire. Looking at it this way it appears that lying to customers is actually good for everyone, and a business manager can keep this in mind when using the deceptive lies and bait to force through a sale.

The most important part of the game matrix used in the last section is reproduced below. It shows the possible payoffs for a buying customer and business if a customer compromises his ideal of getting a high quality product for a very low price (i.e. free), and instead cooperates on the business's terms to end his time consuming search and get the product at an acceptable (but not ideal) price. With a bait and switch a business acts like it's going to compromise on its own incentives to charge an extortionate price, to instead offer value to the customer and make the sale to gain a profitable payoff (5), and the payoff this offers to a buyer (5) is what caused their compromise in the first place. But once the sale is made the seller raises its demands, acting like the sum of money is not enough to get the item that was promised, to screw over the buyer and win the maximum possible payoff (10), forcing a loss (-5) on the poor buyer who fails to get a return for his money.

If there is no obvious way for the buyer to get justice should a seller lie and deceive them, then a business may implement a blatant bait and switch technique. They can promise one thing in return for a customer's money, and simply not make any attempt to deliver what was offered as bait. This phenomenon is often commonplace in tourist traps, as those using the nearby businesses will have little knowledge about the local procedures to achieve justice, and the manipulative seller may hide behind a different language and customs, insisting that his behaviour is perfectly normal in that area. With a steady stream of tourists the seller may not mind that the strategy can only be used once per person, as they will always be able to find more suckers.

Even beyond tourist areas where most people know to keep their guard up, many people can be victim to the tricks of a manipulative business. One popular business strategy is to simply charge a high price for a product. Many people naturally assume that a higher price automatically means a better product, with the phrase 'you get what you pay for' summarizing the idea. If presented with two similar goods it is human nature to assume that the more expensive good is superior in some way, and a buyer could select it over a similar product on this basis alone only to find out that it's no better. If the seller hasn't lied about the details of the good then there's little the buyer could do to get his money back, as the bait and switch was not an active strategy that the business conducted but only existed in the buyer's own subjective opinion, and the authorities can't punish a business because a customer foolishly got their hopes up. As a result the seller has won the game and secured the highest possible payoff in the transaction, while the buyer has lost the game and paid far more than he needed to, for a product that's no different to those available at a far lower price.

In the tourist example where a business takes a customer's money and offers nothing or very little in return, the vulnerable tourist buyer will want their money back, and the deceptive business manager or worker will give something akin to an uncaring 'make me' response, showing a general disinterest in the customer's predicament. And if a business charges a very high price for a product before a customer finds out it's no better than lower price substitutes, the buyer will also insist on a refund and claim deceit, an accusation which the business will respond to with an innocent plea and a 'prove it' attitude. Both the 'make me' and 'prove it' attitudes by a business represent the raising of their demands, and the switch from mutual compromise to one-sided compromise only on the part of the customer.

The success of this bait and switch strategy is based around what is known as the 'big lie' technique. The idea is that while most people have told a lie at some point in their life this is only out of basic necessity, perhaps to spare a loved one's feelings when telling the truth would be cruel, or to cover themselves in times when admitting guilt would incur a harsh punishment. Most people are thought to be honest as default and only lie when they have to, as truth helps them to learn from their mistakes and understand the world around them more effectively.

As a result the average person would expect others to be the same, and only lie to the extent that it was necessary. They may expect a business to cut corners on service effort and charge a price that's a little higher than it should be, as they would too if they had to, but they wouldn't expect a business to charge one hundred times their unit cost, or to not even have the product they claim to offer. This inherent trust allows manipulative sellers to pull their bait and switch techniques on people, and enjoy wealth gains while often offering little or nothing in return.

But after being screwed over enough times people become wary of bait and switch strategies being used against them, and may abandon the idea of maximizing their utility with a hunt for the best products, to instead search out products that are 'good enough' and that they can rely upon. The popularity of discounted price items or big and established brands is a testament to this, as they aren't necessarily the best but customers can be sure the products aren't the worst, and they won't be ripped off.

At a collective level a sense of anger against those who don't follow the rules of the game has increasingly seen authorities clamp down on rogue sellers, and there are now many procedures in place to protect buyers if they feel that their cooperation and compromise hasn't been matched by a business. Popular online sellers such as eBay and Amazon Marketplace offer buyer protection guarantees, where the corporations promise to force through a refund if the buyer is not completely satisfied. But this hasn't put an end to the bait and switch strategy used by manipulative sellers, and it simply redefined the boundaries, forcing sellers to be more careful in hiding the 'switch' part of the routine. Even with the buyer protection programmes offered by the likes of Amazon and eBay, a cunning business can still use a similar strategy against customers to make significant money gains, as will be explained below.

Creating an online business is perhaps the easiest way to create good bait, as buyers would never be able to see the firm face to face, and everything could be controlled and done at the manager's pace. An online business also has the advantage of very low start-up costs, and there's no need to invest in expensive business premises for example. A well-designed site starts things off, while some fake good reviews and positive feedback could be created using additional accounts, either on the website itself or on popular review sites, perhaps going into some depth to make the words appear more realistic.

Prices for the bait-creating business would have to be comparable or a little lower than those of the trusted firms it competes with, otherwise customers would simply go with the more established seller offering the same products. But of course the new business isn't able to match their rival's prices for the in-demand products, as noted above, and this is why they have to work hard to create bait in the first place. However, the bait-creating seller can get around this with a little trick, and although they will use stock photos of the popular in-demand products that buyers want, this is not what they will be selling to their customers. Instead they might give them the same basic product but an inferior and less popular version or brand, one they can get and make a profit on as it was available for stock purchase at a low price. This would of course violate the rules of the likes of eBay or Amazon Marketplace, but the manipulative seller has a plan to get around this.

A buyer will see the photos and description of a quality branded product at a good price, just as the bait-creating seller planned, and then be reassured that the business can be trusted by seeing good reviews testifying as much. The buyer will compromise on his original goal to get the product at a huge discount, and cooperate with the seller by paying the money required to buy the good, thus ending his search for the product. After making the purchase the buyer will feel satisfied, knowing that if anything goes wrong he has the power of eBay or Amazon in his corner to put it right.

Several days after being told the order is dispatched and waiting for it to arrive in the mail, beyond the time such a delivery would normally take, the product hasn't come. If the buyer knew for sure that the business hadn't sent the product, going against their obligation to cooperate, the buyer would take the evidence to eBay or Amazon and force through a refund, but they can't be sure this is what happened. Perhaps it was the postal service that was non-cooperative, and the seller did appear to be a reputable business while any postal delivery service can be unreliable at times. The buyer may look for further evidence one way or another, and find a tracking number provided by the seller, but this leads to nothing as the tracking service suggests the number is wrong. But perhaps the business seller is so successful that they made a mistake in their busyness, and it seems unlikely that anyone would deliberately create a fake tracking number when this would be found out.

The buyer waits a few more days, and might even contact the seller to ask when it was dispatched, and just as it reaches the point where they're about to give up and push for a refund, the item arrives. It almost seems like the business was waiting to see if the customer noticed the product hadn't come before they even dispatched it, but surely a professional business wouldn't do something like that. At this point many people will feel relieved that the product finally arrived and guilty for ever having doubted the seller. And that will be the end of the transaction. Many buyers will not even realize that they haven't achieved the payoff they had hoped, as the item isn't exactly what they ordered. This would be unlikely for higher-priced products or regular purchases, but for rarer buys of required household items a buyer may purchase unknown brands based solely upon reviews and ratings or a sales promotion, and they may not notice that they haven't received what was ordered. Those buyers who do realize it will contact the seller, thinking that it must have been a genuine mistake, and surely no-one would deliberately send the wrong item.

A bait-creating business will respond to a buyer reporting an incorrect product by raising its demands, the predictable 'switch' part of a bait and switch;

'Please send photos of the product you received'

The business is basically demanding proof that the customer isn't just trying to screw them over with lies, a classic case of projection. Many aggrieved customers will give up at this point, as they might be unfamiliar with the process of using a digital camera and sending photos by email, or feel they've done something wrong by expecting more than they received. After all, they did get a product that vaguely represents what they ordered and it only cost a small sum of money, as low cost products are all the business sells for some reason. Those who continue on and send the photos will face more demands by the business;

'It's the same product you ordered'

This is a demand that the customer prove the difference between what they got and what they ordered. If this demand is met the firm may still not do what the customer wants and deserves by law, and instead demands that the buyer consider alternative proposals;

'We'll offer a small discount if you keep the product'

When the buyer makes another response to insist on a refund the seller may demand they incur greater expense;

'Here's our business address, please sent it there and we'll refund the original cost when it arrives'

But there's no mention of the postal charges incurred to return the product, and the seller expects the buyer to pay this significant charge. At this point the only way out for a buyer is to call on eBay or Amazon, who will ultimately force through a refund for the buyer, and the customer will also get to keep the item the seller refuses to accept as a return. Most customers won't want to take it this far, after all they did receive a product of some description, and it's not worth going to war with a seller just because they aren't completely accommodating, just to receive a small sum of money. And they'd either have to keep the wrong item they'd now have received for free, which feels like stealing, or spend time and effort and a sum of their own money to return it, which isn't worth it given the low cost of the product. With the two options they face a buyer may feel the right option is to simply let it go. But they may feel differently if they knew that every single step here was planned out in advance in meticulous detail by the scheming business, which predicted buyers' likely responses and devised a strategy to exploit and beat them.

Most buyers would never think to look again at the tracking number they were provided by the seller. But if they did check the tracking they would see that the item was only dispatched the day before they received it, which explains why the tracking didn't work earlier. It seems the seller had secured a tracking number and then deliberately not sent the item, as they wanted the customer to be more focused on receiving the product than making sure it was the right one, and they also wanted to start a process of self-doubt in the buyer. The business's focus is to convince the buyer that he's the one raising his demands and breaking the mutual compromise agreement, and overstepping the rules of the game, and that they're a reputable business that just wants to cooperate. All of this is designed to stop the buyer invoking their ace card, and calling on the likes of eBay or Amazon who will always see things from the buyer's point of view.

The outcome a customer faces may not always have been planned out in advance in an adverse selection scenario as just explained. It may be a case of moral hazard where a business simply takes the most productive short-run option available as the situation unfolds. But either way it makes little difference and the bad result is exactly the same for the consumer.

In times where a seller's behaviour is carefully watched by authorities the bait and switch strategy simply becomes more indirect. With a business unable to make the first move and openly switch to a role where they screw over the buyer, they have to get the buyer to invite and allow the raised demands by pushing toward greater communication. A business may achieve this by deliberately causing problems that need to be worked out through further interaction.

Repeated Gains

Although it may be highly annoying to the average decent person a bait and switch strategy can be successful for a business no matter what safeguards are put in place in an attempt to prevent it. When one side is dedicated to a specific strategy and the other side is trusting and naive it's little surprise who wins.

However, as noted earlier a bait and switch is only likely to work once per customer, as even though a buyer may not figure out the depths of a business's manipulative cunning, they know when an interaction didn't go exactly as they wanted and they'll look elsewhere next time. The strategy of bait and switch can therefore be a lot of work as a business will need to constantly find new suckers, and their act is based around doing one thing and convincing potential buyers that they're doing the exact opposite. Because of this a bait and switch strategy has severe limitations, and while it can earn a bit of money, it's unlikely to generate large amounts of wealth for a business.

The most successful businesses look beyond potentially lucrative but less frequent one-off transactions just like an efficient customer does, and instead seek to build a long-term business relationship, to gain the tempting long-run bait of repeated payoffs. This is instinctively appealing for a firm, even though the idea may initially appear counterintuitive as the modified business-customer interaction below suggests.

	Buyer compromises	Buyer raises demands
Seller compromises	1 1	10 -5
Seller raises demands	-5 10	0 0

In this example the payoffs from cooperation are much lower for both parties, as a seller will only make a very small profit (payoff of 1) if it makes a mutual compromise on price with the buyer, to allow their valuations to meet and facilitate the sale. But the gains from successfully screwing over the other side remain just as high as before (payoff of 10), now at ten times the compromise payoffs. Unless a business believes there will be ten or more separate interactions with the other (10 x 1 = overall payoff of 10), it appears to be in their best interests to try and screw over buyers, and these payoff incentives should push them to raise their demands.

But in many real world scenarios the gains from exploiting a customer in a business transaction vastly exceed the payoffs on offer from cooperation, yet a business doesn't raise its demands to take advantage and instead cooperates and compromises to meet their customers' needs. This behaviour holds even in scenarios where the business knows with near certainty that there won't be many transactions, and not enough for the sum of cooperative transactions to offer a higher total payoff than a single successful manipulation could.

It may make sense for a seller of luxury goods to offer low prices as bait to draw customers in, prices the firm will then be legally tied down to, but once the customer is there the luxury seller would be expected to try and rip them off anyway they can. Continuing to cooperate with good deals might only offer a very low payoff (e.g. 1), while successfully manipulating buyers into expensive additional purchases they don't need may give a very high financial payoff (e.g. 10). With the luxury business knowing that the average customer is highly unlikely to make ten return visits continued cooperation and compromise may appear surprising. But this ignores the fact that the compromise or raise demands game here will be repeated for every single round of future interaction, and both sides will have the option to change strategy in future rounds.

The following diagram is a game tree from a business seller's point of view, and this shows how a business-customer relationship can play out over time. Taking the same payoffs as seen earlier in the previous game it uses forward induction to look beyond the one-off transaction represented by a game matrix, to reveal how a past choice can affect future options. This shows the incentives a player has in the present if the long-run (LR) is taken into account.

Business's payoffs in a LR game with a buyer

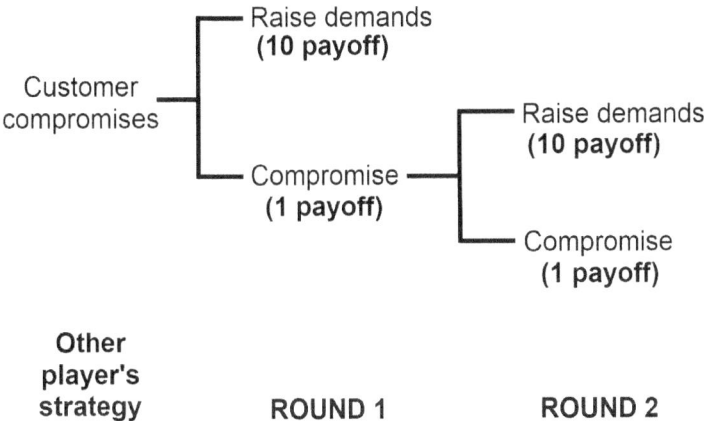

For a productive and appealing transaction to occur a business first needs a customer to compromise, as noted earlier. Once this happens a business can raise their demands with a bait and switch as described in the last section, but although this gives a payoff of 10 the customer will avoid any future interaction and that's all the firm will get. If a business instead copies the customer's compromise then they will get a payoff of 1, and with a buyer not scared away by being screwed over further rounds of interaction are possible in the future. A business could cooperate again for another payoff of 1, or decide that now is the time to screw over the buyer and secure the payoff of 10, raising its demands to manipulate the customer into handing over a larger sum of money that was required, or simply not delivering on its promise to offer value in return for the buyer's cash. Even just one round of compromise by the business could offer a higher total payoff ($1 + 10 = 11$) than betraying the buyer from the start. And two rounds of cooperation could offer even more ($1 + 1 + 10 = 12$), and so on.

But the benefits a business gains from long-run cooperation and compromise are dependent upon a customer's continued compromise, and if the customer doesn't give this then it may be best for a business to avoid giving it too. It may seem a stretch to assume that the customer will keep cooperating, and not take the opportunity to either attempt to screw the seller over by raising their own demands, or simply avoid any further interactions with the business as they only ever required one product. But the customer's own incentives are very similar if not identical to those of a business, as the game tree from the customer's point of view shows.

If a business compromises with a customer, a prerequisite for an appealing interaction for the buyer, then a customer could take back their own cooperation and raise their demands, perhaps finding some flaw with what the business has offered and demanding a lower price to make up for it. This would give a payoff of 10 in this example, but it would see the seller put the customer on their blacklist, and vow to never do business with them ever again.

Customer's payoffs in a LR game with a seller

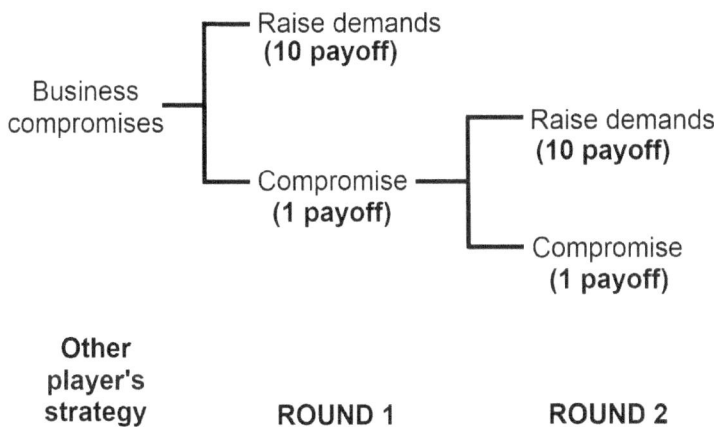

On the other hand a customer could maintain their cooperation and match the business's own compromise, to earn a payoff of 1 and leave future interactions open. The next time the two sides meet the customer could cooperate again for a payoff of 1, or take the opportunity to try and screw over the business by raising their demands, for a lucrative payoff of 10. One round of compromise followed by a round of raised demands would earn a superior payoff for the customer (1 + 10 = 11), with further rounds of cooperation adding to the total payoff.

While the benefits of cooperation are entirely dependent on the other side following the example, both a business and a customer have the same incentives for maintaining cooperation to gain higher individual long-run payoffs, and this makes it more likely that both sides will return for further interactions if the other side cooperates. It's also not necessarily a case of cooperating for a few rounds and then betraying the other side before they have a chance to do it to you first. If a business was to screw over the seller on the third round of interaction after two rounds of cooperation their payoff would be 12 here (1 + 1 + 10), but this represents a loss compared to the higher return of 13 (1 + 1 + 1 + 10) they could have achieved if they'd waited for one more round. The same result holds for the customer, as one more round of interaction is always best for both parties.

In the example here the payoff from exploiting the other side of the business transaction is ten times the payoff linked with mutual cooperation, and even in this situation there are gains to be had from compromise. Yet in practice the difference between the two payoffs may not be as significant, and this would make compromise even more rewarding for all those involved.

Even if a business or customer is almost certain they'll never see the other side ever again, for example if the business knows the customer is a tourist in an unpopular resort, or if the buyer knows that they'll never buy another house after the one they've just acquired, it still pays to cooperate. They might be wrong that they'll never meet again, and even if not, that buyer or seller will know other people who might be turned into payoff opportunities. If the tourist business was manipulative the customer will spread the word, and that's less business of any kind in the future. But if they cooperated they'll receive a good word and the customer's friends may soon pay a visit, allowing the business to either take advantage of them or cooperate for greater potential gains in the future. The same is true for a seller, and a good customer or client may be rewarded by the seller giving access to his friends in other fields, which the customer could then treat as they wish.

Last but not least, the actual process of raising demands and being successful is not a given, or everyone would do it every time. But a customer who cooperates first will have time to learn more about a business which they could use to find any possible weaknesses to exploit, while a business will build trust by cooperating first and this can be called on later. The most popular brands can now get away with charging high prices for their goods, and can even create terrible products and still have consumers buy them, all because they have now built blind trust though cooperation with customers in the past.

Overall the incentive for a buyer or seller to not exploit the other side and instead cooperate could be seen as chasing specific long-run bait; the bait of repeated gains and greater opportunities to exploit the other side in the future.

The Business of Show

The last two sections have explained the strategies a business may use for the goal of wealth generation. A bait and switch technique can potentially be successful in the short-run, generating moderate wealth payoffs from exploiting the naivety and general decency of honest buyers. To gain larger amounts of wealth and longer-run payoff returns a business can commit to a path of compromise and cooperation. This can draw in customers with the bait of both low but repeated payoffs at first, and also the ongoing potential for a maximum one-off payoff, if the buyer was able to successfully manipulate the business in a future round of interaction.

Both short-run and long-run strategies to persuade a buyer to spend their money are based entirely around the idea of bait. A buyer will happily give a one-off sum of money for the bait part of a bait and switch, even though it's not based upon anything of substance, but they'll want the money back when faced with the switch. And it's not gaining a high payoff that motivates buyers to cooperate in the long-run or they would push for that right away, it's the bait of future payoffs, which may or may not be available depending on the business's future behaviour. In simple terms it's not the facts or reality of the payoffs on offer that gets buyers to pay money, but the bait or fantasy of them. A smart business will therefore try to appeal to buyers' fantasies wherever possible, and although individual buyers' specific preferences may differ, all of them can be relied upon to prefer certain features. For example, all customers would like to deal with a business seller that cares about their needs, and it's in a seller's best interests to show signs that they are a caring firm.

While appealing to buyers' fantasies will draw in buyers and their money, a firm using a bait and switch strategy will have to face some buyers who are outraged at the switch, successfully forcing through refunds to take back some of the firm's acquired wealth. And a firm using cooperative long-run bait to generate wealth may find it less successful than expected, as some customers lose faith that the future payoffs will ever materialize and take their money elsewhere. The problem for a business is it typically can't just rely on bait and fantasy, but has to deliver a good or service based on substance at some point.

Show business is a type of business unlike any other, and in this field the business doesn't have to promise a specific physical good, digital product, or a customer-focussed service. Yet despite not offering any of the features normally associated with a firm, some of those involved in the show business field are among the wealthiest people in the world. Actors and other entertainers do nothing more than sell a fantasy world of make believe, and people worldwide line up in their millions to hand over their money and complement those responsible for a great show. And because each act is based entirely upon the creativity of the entertainers who 'show' the buyer the best fantasy they can come up with, these businesses are not about a seller meeting objective standards that satisfy the buyer, but offering their own subjective product that a paying customer can then select or reject. The business is free to focus entirely on manipulating buyers with bait and fantasy, the most reliable way to make money, and the only complaint a buyer can make is that the bait wasn't convincing and manipulative enough to make them fall for the act.

The subjective nature of show business creates a different game to that usually associated with a buyer-seller interaction. While the two strategies available remain very similar, as 'compromise' or 'won't compromise' (essentially the same as 'raise demands' used earlier), the payoffs are different as shown below.

	Buyer compromises	Buyer won't compromise
Seller compromises	5 5	10 -5
Seller won't compromise	5 10	0 0

One payoff has changed from the familiar prisoners' dilemma game used to describe earlier buyer-seller interactions, and the buyer's payoff for compromising with a non-comprising seller has changed from a -5 loss to a gain of 5. As before if neither party compromises then there's no interaction, for payoffs of 0 each, and if both compromise the buyer gains a product (entertainment here) of some value while the seller gains some money, for payoffs of 5 each. In a scenario where the seller compromises and the buyer won't, in the top right of the grid, the buyer gains a 10 payoff by paying less money than the product's worth, while the seller suffers a -5 payoff loss for going along with this.

But the changed payoff in the bottom left of the grid, where the seller won't compromise (payoff 10) and the buyer does compromise (payoff 5), is more complicated than before. In the past it was all about a seller gaining by getting far more money than his minimum selling price, while the buyer loses out (for the former payoff of -5) by paying more than was required for a product that wasn't worth it. But this is based around the idea of objective value and facts, and as noted above show business is based on subjective value and manipulating emotions.

The 10 payoff a seller gains from not compromising isn't just based around making more money here, but also being able to create the entertainment product they want, as opposed to being forced to follow someone else's ideas which may be difficult to adapt to and convert into a good show. The buyer's 5 payoff is because although it might cost a little more money, dragging the payoff down, the only way they can gain an entertainment product they value is to give up on their own ideas for the show (i.e. compromise), and let the talented seller express their subjective ideas and do their thing (i.e. don't compromise). There's no way around this as the entire purpose of show business is escapism from objective facts and depressing everyday reality, and clearly the buyer can't do this alone or they wouldn't even consider entertainment, so they turn to someone else to manipulate their emotional triggers and offer a high payoff.

With this change made to the original matrix it's worth running through the buyer-seller game again, to determine a player's best response to a strategy from the other player. First the game is examined for the show business seller.

	Buyer compromises
Seller compromises	5
Seller won't compromise	10

If the buyer compromises then the seller shouldn't, as avoiding a compromise would earn a higher payoff of 10 in place of a payoff of 5.

	Buyer won't compromise
Seller compromises	-5
Seller won't compromise	0

And if the buyer won't compromise the seller again shouldn't compromise, to avoid a -5 negative payoff and earn a 0 payoff instead, ending up no worse than he started.

Overall the seller's best response is to not compromise no matter what the buyer does. This strategy dominates the compromise option, always earning a better payoff and never a worse payoff to make it the buyer's dominant strategy.

Next the game is examined for the buyer of the entertainment product, who faces different payoffs to the seller.

	Buyer compromises	Buyer won't compromise
Seller compromises	5	10

If the seller compromises then the buyer should do the opposite, and a decision that they won't compromise earns the buyer a superior payoff of 10 instead of the 5 on offer with cooperation.

	Buyer compromises	Buyer won't compromise
Seller won't compromise	5	0

And if the seller won't compromise the seller should again do the opposite, compromising for a positive payoff of 5 compared to no return at all and a 0 payoff if he copies the seller's action.

Overall the buyer's best response is to do exactly the opposite of whatever strategy the seller follows. This information can be combined with the seller's dominant strategy to not compromise noted above to predict the result of this game. The seller can be expected to never compromise, while the buyer does the opposite of what the seller does. This should lead to the Nash equilibrium expected result underlined above, where the seller won't compromise but the buyer does, as he follows the opposite strategy to the business. This gives the seller their highest possible payoff of 10 here, and earns the buyer a satisfactory payoff of 5.

	Buyer compromises	Buyer won't compromise
Seller compromises	5 5	10 -5
Seller won't compromise	<u>5</u> <u>10</u>	0 0

As with all Nash equilibrium results this outcome is stable, and neither player can achieve a better payoff by changing their strategy without the other player also changing their own strategy. Any attempt to single-handedly change strategy will see a player worse off than before, as the buyer's payoff will fall from 5 to 0, while a business's payoff will fall from 10 to 5.

This representative game matrix highlights why show business and entertainment can be so lucrative for a seller. It may be one of the few types of game where non-compromise and putting themselves first is a stable Nash equilibrium for a business. The seller simply creates a product without compromising on their subjective ideas, and waits for buyers to notice them and realize that they'd be better off compromising on their own ideas of entertainment to follow those offered by the seller, at a price of course.

While those in show business can get away with nothing but bait based around their own fantasy world, society and civilization can't function on entertainers alone, and businesses offering goods and services based on substance are needed. But firms in other sectors can still learn something from those in the business of show. They are free to build a fantasy world for their products, based around advertising and marketing used in any and every way a firm can think up. A firm could show a magazine ad, billboard, or commercial with actors paid to smile as they sample the firm's product, to build the fantasy that the product will bring happiness, wealth or success. Or the firm could use a name and logos that suggest the product is fun, stylish, clever, or just simply better, by showing traditional indicators of these factors in their products to draw unconscious links. A business needs to associate its goods with positive emotions, and although a buyer may be disappointed with the product itself, if the buyer irrationally associates it with hard to get feelings of happiness and success they'll believe the problem must be them and not the product, and the sale won't be threatened.

Focusing on buyers' emotions instead of just facts can support a seller's goal to increase sales and profits, but even if this is done well there are still other concerns that may provide more of a challenge. There are always going to be other firms doing the exact same thing and competing for the same customers, and the challenges posed by other businesses is the focus of the next section.

Business Competition

So far the focus has been upon how a business can attract buyers and gain wealth, with the only obstacle the customers' demand for value. But in real life there are other challenges too, and even if a business knows how to attract buyers they still face competition from others with the same knowledge. A seller's main challenge will not necessarily come from customers but from rival businesses, which seek to usurp their role and claim buyers' money all for themselves.

A firm that stands as the only business in its sector is a monopoly, and with no competition it can set whatever price it wants. Its choice will see a relatively high monopoly price to maximize profit, keeping the price below a level that would repel buyers but ensuring a high return per sale. However, a firm is unlikely to find itself alone in a sector, and there will almost certainly be other firms competing for the same customers with a substitute product or service. In order to take sales away from business rivals a firm will seek to lower prices and undercut the competition, as this is the quickest and easiest way to increase the relative value of its products and make them more appealing to prospective buyers.

The game matrix below examines a duopoly, a two firm industry, representing the interaction between the two as a game. Each can either follow a strategy to select a relatively high monopoly price for their products, or a lower more competitive price, and the matrix gives the corresponding payoffs for the possible scenarios that will result.

	Firm 2 monopoly price	Firm 2 competitive price
Firm 1 monopoly price	5 5	7 1
Firm 1 competitive price	1 7	3 3

If a firm was the only seller in an industry it would select a profit maximizing monopoly price for its products, gaining monopoly profits for a possible payoff of 10 (on a scale of 0 to 10), and in the duopoly here if both firms select a monopoly price they will share these sales and profits for payoffs of 5 each. If one firm selects the monopoly price and the rival undercuts it with a more competitive price, then the competitive firm has a lower profit per sale but takes the majority of the other firm's sales, for a payoff of 7. Meanwhile the firm going with a monopoly price like it's the only firm in town loses most of its sales, but it still gains a payoff of 1 as there will be a minority people who actually prefer to pay more for a product, naively assuming that a greater price always means greater quality. Finally, if both firms select a competitive price then they each gain payoffs of 3, sharing the total profits and payoff linked to that price.

To predict which of the two prices each firm will select a small part of the game matrix can be isolated, to reveal a firm's best response to the strategy chosen by the other player. The game segment below shows firm one's options and corresponding payoffs given a choice by firm two to select a monopoly price.

	Firm 2 monopoly price
Firm 1 monopoly price	5
Firm 1 competitive price	7

If firm two selects a high monopoly price then it's best for firm one to go for a lower and more competitive price. This offers a higher payoff of 7 compared to 5, as the undercutting firm's lower price gains the majority of its rival's customers and sales.

	Firm 2 competitive price
Firm 1 monopoly price	1
Firm 1 competitive price	3

And if firm two chooses a competitive price then firm one's best response is again to select a competitive price. This gives a payoff of 3 as the two firms share the market's sales between them, while selecting a higher monopoly price would only offer a payoff of 1 as firm two's lower price would steal sales and profits.

Whatever firm two does it's best for firm one to select a competitive price over a monopoly price, and because payoffs are the same for both firms in this game the result also holds for firm two. The strategy to pick a competitive price dominates the strategy to go for a higher monopoly price, and it's therefore the expected strategy that both firms will follow. This gives a Nash equilibrium and predicted outcome where the two duopoly firms price competitively, giving payoffs of 3 each as underlined below.

	Firm 2 monopoly price	Firm 2 competitive price
Firm 1 monopoly price	5 5	7 1
Firm 1 competitive price	1 7	<u>3</u> <u>3</u>

But as the matrix shows, the mutual payoffs of 3 could have been bettered, and if both firms had collectively selected a monopoly price they would have achieved joint profit maximization and superior payoffs of 5 each. This would be the Pareto efficient outcome for the two firms, and it would have seen them share the maximum possible business profit of 10 evenly between them, which the other options all miss out on.

With the individual incentives of the two firms pushing them away from a mutually beneficial outcome and toward damaging competition, it's worth investigating how far this process might go. A monopoly price can offer a high profit but may be easily undercut by a rival, while a lower competitive price offers some profits but they're sufficiently low that such a price is hard for a business rival to undercut, at least not without giving up on its own profits. But in a 'price war' scenario a firm is not concerned about making a profit, only gaining market share and destroying its rival. A firm could set a very low price that covers their ongoing costs but offers negligible to no profit at all, with prices going as low as to equal marginal cost, $P = MC$. This price war strategy would be designed to reduce a rival's sales and payoffs and remove the business competitor from the industry, to allow the remaining firm to set a profit maximizing monopoly price with no-one to stop it. The game matrix below shows a game between the two firms in the duopoly once again, but this time the choice of strategies is between staying at a competitive price or cutting them further in a price war.

	Firm 2 competitive price	Firm 2 price war
Firm 1 competitive price	3 3	0.5 1
Firm 1 price war	1 0.5	0 0

The payoffs where both firms select a competitive price is the same as before, giving a payoff of 3 to each firm, while if both firms push for a price war they will both price at their marginal cost, and receive zero profits and a payoff of 0 each. If one firm pushes for a price war while the other has a higher but competitive price as before, then the price war firm takes most of its rivals sales but only earns a payoff of 0.5 (on a scale of 0 to 10). One firm individually pushing a price war firm may earn a little profit at first, taking sales away from its higher priced rival with a declining price, but if the focus is not their own profits but to take down those of a rival the price war will soon see its price fall equal to marginal cost. And at this point no further profit can be made no matter how many sales the firm makes. On the other hand, the firm sticking to a competitive price while a rival pushes a price war will get a payoff of 1, and even though its price is higher it will still get a few sales, as some buyers will prefer a higher priced product and assume it must be of higher quality.

To find the result of this game a small part of the complete matrix is again isolated, to show a firm's best response given a certain strategy by the other firm. First firm one's best response will be investigated in a scenario where firm two selects a competitive price for their products.

It's in firm one's best interests to match any competitive price selected by firm two. This will offer an acceptable payoff of 3, while an attempt to start a price war by undercutting the rival's price will only give a reduced payoff of 0.5, because the firm's products will ultimately not be priced high enough to generate a clear profit.

	Firm 2 competitive price
Firm 1 competitive price	3
Firm 1 price war	0.5

If firm two pushes a price war then it's in firm one's best interests to avoid falling into the trap of following suit, and they should instead price competitively. This will offer a low but positive payoff of 1 as some buyers will happily pay for a higher priced product, while a price war would see price set equal to marginal cost and no profit at all for either firm, giving a payoff of 0 for firm one.

	Firm 2 price war
Firm 1 competitive price	1
Firm 1 price war	0

If firm two pushes a price war then it's in firm one's best interests to avoid falling into the trap of following suit, and they should instead price competitively. This will offer a low but positive payoff of 1 as some buyers will happily pay for a higher priced product, while a price war would see price set equal to marginal cost and no profit at all for either firm, giving a payoff of 0 for firm one.

Putting the results together reveals that selecting a competitive price dominates the strategy to push a price war, and it always offers a better payoff for firm one and never a worse payoff. With identical payoffs for firm one and two this result is also true for the second player in this game, and this gives a Nash equilibrium and expected outcome where both firms price competitively, with payoffs of 3 each.

	Firm 2 competitive price	Firm 2 price war
Firm 1 competitive price	<u>3</u> <u>3</u>	0.5 1
Firm 1 price war	1 0.5	0 0

It seems that although firms in a multi-firm industry have an individual incentive to undercut the joint profit maximization price, where each firms selects a monopoly price, they don't have an incentive to lower prices below a competitive level and deliberately push for a price war that would threaten their existence. However, there's always the risk that firms may engage in competitive pricing that gradually leads to lower price war prices without the firms realizing it. While they may not choose a price war firms may find themselves forced into one.

Although the analysis here predicts an outcome of competitive pricing among firms this will not necessarily hold in all circumstances, as businesses will be unsatisfied to miss out on higher prices and profits, and they'll seek out ways to move toward a monopoly price to achieve profit maximization. The next section looks into the prospects for this goal and how it may come about.

Tacit Collusion

The last section concluded that two firms in a duopoly industry would price their products competitively, giving a result of competitive prices across the industry. This sees both firms miss out on the superior joint profit maximization outcome, as their individual profit seeking and desire to avoid being undercut in price by their rival makes each avoid a more lucrative monopoly price.

	Firm 2 monopoly price	Firm 2 competitive price
Firm 1 monopoly price	5 5	7 1
Firm 1 competitive price	1 7	<u>3</u> <u>3</u>

This situation is essentially a prisoners' dilemma game, where both firms would gain higher payoffs from cooperation, but they can't find a way to enforce this in a world where they're in direct competition. In the business world another factor also comes into play, as explicit collusion to maintain high prices is illegal and will be clamped down upon by authorities.

Earlier on it was noted that in a repeated game businesses and customers may overcome their short-run incentives to betray the other, by focusing on the greater gains on offer from extended long-run cooperation. This cooperation was not dependent on players getting together to make an agreement, but on understanding their own individual incentives over the long-run and the similar incentives facing the other player in the game.

Two firms selecting their prices in a duopoly are also involved in a repeated game, lasting as long as the two firms remain in the industry. And the two firms don't need to attempt to make an explicit (and illegal) agreement for price collusion, because both will naturally understand their own payoffs and the incentives at play, and a cooperative outcome could come about with unspoken but knowledge based tacit collusion. Firm one could individually select a monopoly price for its products, as is in its own best long-run interests, and then see if firm two made a (relatively quick) individual choice to do the same, knowing that it also has the long-run incentives to do so. If firm two did this and stuck with it then both firms could maintain monopoly prices, enjoying joint profit maximization payoffs for the indefinite future. And if one firm ever chose to price at a lower competitive level for the short-run gains then it would be clear that joint profit maximization can't be relied upon, and the other firm would commit to competitive pricing from then on, irrespective of whatever excuses or apologies the 'cheating' firm made. Although they may not be able to walk away from each other when the other side breaks from cooperation, like a business and customer can, two firms in a duopoly can still have a zero tolerance policy.

The game tree below shows the strategies and corresponding payoffs available to duopoly firm one over time, assuming that firm two selects a monopoly price. It uses forward induction to see how a past choice affects those available in the future.

LR payoffs if a rival selects a monopoly price

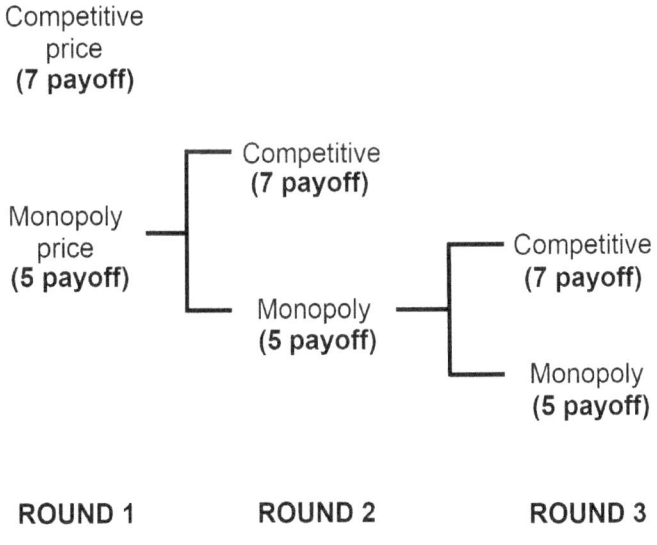

If firm two selects a monopoly price then firm one can either do the same or choose a lower more competitive price. If it selects a monopoly price like the other firm then the game tree branches extend into another round, offering another chance of 5 or 7 payoffs. But if firm one ever selects a competitive price then it gains a higher payoff of 7 compared to 5 in that round but the payoffs from thereafter transform into those below, as firm two will respond by always selecting a competitive price from then onwards.

LR payoffs if a rival selects a competitive price

Competitive price **(3 payoff)**	Competitive price **(3 payoff)**	Competitive price **(3 payoff)**
Monopoly price **(1 payoff)**	Monopoly price **(1 payoff)**	Monopoly price **(1 payoff)**
ROUND 1	**ROUND 2**	**ROUND 3**

If the other firm selects a competitive price for its products then firm one can either do the same for a payoff of 3, or go with a higher monopoly price for a payoff of 1. This choice will not affect the options in future rounds as firm two has already chosen to play the competitive price at every future opportunity, and firm one will again face the choice of a competitive or monopoly price, with corresponding payoffs of 3 and 1 respectively. It can be expected to select a competitive price in every round for the superior payoff this strategy offers.

With these representative payoffs for firm one a prediction can be made over the strategy it will follow, and because payoffs are the same for both duopoly firms the result will also hold for firm two. If firm one selects a competitive price in round one it gains a payoff of 7 against the other firm's monopoly price, and then a 3 payoff for every round after as both firms price competitively. After three rounds it would have (7 + 3 + 3) a total payoff of 13.

If the defection from a monopoly price came in round two instead, then following three rounds firm one would gain (5 + 7 + 3), and a total payoff of 15, which is greater than before. And in a scenario where firm one abandons a monopoly price in round three, then it would gain the payoffs from two rounds of joint profit maximization before the highest individual payoff available in the game, and these payoffs (5 + 5 + 7) give an overall payoff of 17.

The specific payoffs and profit sizes may differ in reality of course and the relative difference between them may be completely different, but this result shows the general idea that the longer a firm cooperates the greater its total payoff is likely to be.

Unfortunately for the two duopoly firms individual incentives may still prevent them from sharing the profit maximization result, despite their committed (but unofficial to keep it legal) monopoly pricing strategy, and their zero tolerance for a 'cheating' move by the other firm to undercut prices. The problem is that it's only the threat of a rival's punishment keeping the two firms from acting on their individual incentives to lower prices, as unlike a buyer-customer game seen earlier the firm here isn't at risk of getting a bad reputation with customers, who would naturally approve of a price cut. And when the threat of punishment falls apart in the very last round of the game where there can be no prospect for future punishment, as there wouldn't be any further rounds for a firm to respond if the other prices competitively, all barriers against acting on their individual incentives disappear. As a result both firms can be expected to select lower competitive prices for their products in the final round of the game. And this can have big implications for every round before it.

When a player in a game knows what will happen at some point in the future they can determine its impact on the present, using a process known as backward induction. Each of the two firms in the duopoly can predict the behaviour of the other in the final round of the game, and this knowledge allows them to forecast their own best response in the round before, which can then be used to predict their best move in the round before that, and so on. In the last round of the game both firms will price competitively (C) due to no risk of future punishment:

Last Round
Firm 1 price = C
Firm 2 price = C

With the result of the last round known, the last - 1 round just before it becomes the final interaction in the game requiring players to devise a strategy, choosing between a monopoly price (M) and a competitive price (C) for their products. Each firm knows that the round after will see the other firm price competitively, essentially implementing a 'punishment' strategy whatever they do. And if they can't avoid this punishment in the next round the firms are not playing a long-run game with repeated choices but a one-off game followed by unavoidable future punishment, and as seen earlier their one-off game payoff incentives suggest they should always select a competitive price. This sees both firms again choose competitive prices in the last - 1 round:

Last - 1 Round
Firm 1 price = C
Firm 2 price = C

With this information known the third round from the end of the interaction, the last - 2 round, becomes the final phase of the game needing a decision on strategy. And the factors just noted come into play again as both firms know the other is sure to select a competitive price in the following round, and they have no reason to select a monopoly price of their own to try to avoid it through cooperation, and they will instead go with the competitive price:

Last - 2 Round
Firm 1 price = C
Firm 2 price = C

This pattern can be calculated back one round at a time all the way to the start of the game, and the outcome will remain the same throughout. A result of mutual competitive pricing may therefore be forecast for the whole game, and tacit collusion appears to be elusive for the two duopoly firms. The joint profit maximization payoffs given above for mutual monopoly pricing over the long-run are accurate, but the firms get greedy and aren't happy with simply getting superior payoffs, instead trying to get the absolute maximum possible overall payoff by only playing a monopoly price until the final round, where they undercut their rival to gain the highest single payoff too.

Even if they know the other firm will do the same it remains the best move, as setting a competitive price always dominates a monopoly price as an individual one-off game strategy. This has the effect of removing the final round from the game as the result is known, and the second last round becomes the real final round, and so on all the way back to the start of the game.

However, despite this theoretical objection there may still be hope for monopoly profits and joint profit maximization between the two firms. The backward induction process just examined depends upon knowledge of the timing of the end of a game, and if the two firms don't know this then they can't work backwards in steps to the present to ruin all hopes of tacit collusion. Yet the two firms can't know when the game will end, as this isn't a direct interaction that they could control such as between a business and customer, but a larger interaction where they are locked in competition and can't escape.

Firms that share an industry can't even be sure that the game will end at all, and the firms could remain competitors in the same industry decades into the future, even after the deaths of the current managers deciding on today's pricing strategy. Players won't prepare to select a lower price in the final round if they don't think it'll ever happen, and without this last round result the earlier rounds won't necessarily see a reduced price via backward induction, and joint profit maximization is possible.

Even if players aren't focused on gaining a superior 'cheat' payoff from undercutting their rival in the last round of the game, they may still ruin tacit collusion by doing this in the very first round based upon forward induction. This could happen if a firm believed the gains from doing so outweighed the future payoffs the move would sacrifice, and this would depend upon the discount rate for expected future payoffs. To be completely accurate all payoffs occurring in a future time should be discounted due to the time value of money, where a sum of money (payoff) is worth more if received today than if the same amount is received in the future. While this isn't really possible with business-consumer interactions which will occur in unknown future time periods, it may be possible with business competitors that a firm knows it will interact with periodically.

Factors such as risk and inflation reduce the nominal value of future income, while interest rates can raise the nominal value of current income if it was invested in a bank, and a discounting rate is needed to account for all of these variables. The more significant these factors are the greater the discount rate will need to be, in order to find the present value of future money income. The calculated discount rate (R) is put into the formula: $1 / (1 + R)^N$, where N is the number of rounds into the future a specific payoff lies. This finds the discounting factor to be multiplied by each future payoff, to give its present value in today's money. If the discount rate R is sufficiently low then the discounting factor will remain close to one and the repeated future gains from tacit collusion, 5 payoff each round instead of 3 in the game above, will outweigh the immediate one-off gain from defecting that doesn't get discounted, 7 payoff instead of 5 here, to support long-run cooperation and monopoly profits. But if the discount rate is high then tacit collusion may prove elusive.

Entry Deterrence

In a long-run duopoly game the two firms involved may be able to achieve tacit collusion and joint profit maximization, overcoming their individual incentives to lower prices to outmanoeuvre the other. This unstated tacit collusion would be based around the shared understanding that each firm would prefer to set a monopoly price for higher long-run profits, and they will only lower their prices if the rival forces it by lowering their own prices and forcing competition.

Unfortunately for firms an industry structure will not stay the same forever, and if the two duopoly firms achieve monopoly payoffs and joint profit maximization then others outside the industry will look on enviously. They will want a piece of the pie and plan to enter the industry themselves, aiming to get a share of the lucrative profits that appear to be on offer.

The two firms in the duopoly will be desperate to keep other firms out, as tacit collusion to keep prices and profits high will be harder and less likely with more firms, and more game players to keep in line. Even if the potential entrants also set their prices at a monopoly level instead of forcing competitive prices across the industry, there will still be consequences for the incumbent firms (those firms which already hold a position in the industry). Three (or more) firms achieving joint profit maximization means less for the original two established firm incumbents, as buyers won't be spending more money yet the profits will have to be shared between another firm.

There are some barriers to entry that can support the incumbent firms already in an industry against newcomers, such as regulatory barriers and customer switching costs, while the industry can also require economies of scale or high unrecoverable sunk costs for a firm to get up and running. But while this can give incumbent firms a 'first mover advantage' over those who aspire to join their industry, this may not be enough to keep out all potential entrants, especially those larger and wealthy firms who already have success in a different business area and feel confident they could excel in another.

The following game tree below shows the game that will take place between a representative incumbent in an industry and a potential entrant, and it looks bad for the incumbent. The established incumbent firm can't do anything to stop the new firm entering their territory in this game, and it's the entrant who makes the first move, deciding if they should go in or stay out of the industry.

Underlined choices and payoffs in the game tree refer to the incumbent, while those not underlined apply to the entrant. This shows that if the potential entrant stays out of the industry they'll gain a payoff of 0, as they've neither gained nor lost anything. Meanwhile the incumbent will gain a payoff of value Vm, where the 'm' represents monopoly profits from joint profit maximization, assuming there's either only one incumbent, or there's tacit collusion to keep prices high if there are several incumbent firms.

An entry deterrence game

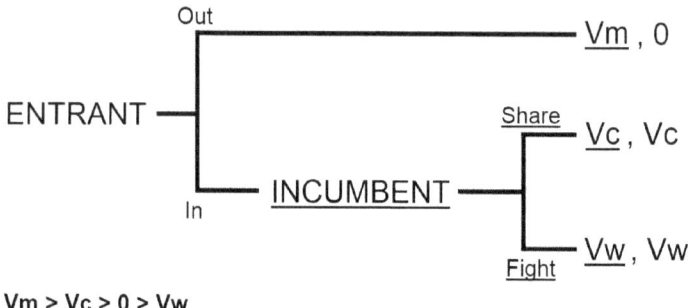

Vm > Vc > 0 > Vw

With the payoff for not entering the industry equal to zero, all it takes for the firm to enter is for the corresponding return to be greater than zero. Their precise payoff will depend on the response of the incumbent firm, who may either fight to keep them out of the sector, or accept their presence without a fight and share the industry with them. If the incumbent shares the industry then both gain payoffs of value Vc where the 'c' represents cooperative profits, which will be above zero for both parties, but naturally less than the profit maximization payoff Vm for the incumbent. And if the incumbent instead fights with the entrant, engaging in a price war where price is set at marginal cost in an attempt to wear down the newcomer through attrition, then both firms will gain a payoff of value Vw, where 'w' represents war. This will be below zero and cause a loss for the entrant who will have spent sunk costs to get into the industry, but now must engage in an unprofitable price war just to stay in the sector. It will also be negative for the incumbent as they won't gain profit during the price war, yet sacrifice the monopoly profits they could have earned over the period.

A potential entrant knows that if the incumbent were to share the industry with them they would gain a positive payoff, with Vc > 0, but if they faced a fight there would be a negative payoff, Vw < 0, less than the 0 payoff on offer if they stay out of the industry. Therefore the new firm will enter the industry if they think the incumbent will share it, and avoid entry if they believe the incumbent will fight instead. Unfortunately for the incumbent the new firm on the scene is likely to know of their own incentives, and that it's in the best interests of the incumbent to cooperate and share the industry instead of engaging in a costly price war. As a result the new firm will enter the industry, and the incumbent will share industry profits.

In a basic one-off game of entry the expected outcome is for an entrant to enter and for the incumbent to share the industry with them. But there are several extensions to this model which can let a firm exercise entry deterrence.

Repeated entry may be a more realistic situation than a one-off game, and it's likely that a regular stream of firms will seek to gain entry into an industry that offers profits. In this scenario it may be a mistake for an incumbent to share the industry with entrants, but fighting them with a price war could help the incumbent gain an aggressive fighting reputation, to deter future entrants who would want to avoid a costly fight.

But fighting entrants is only be a perfect equilibrium strategy if entry is repeated indefinitely, as with limited entry prospects in the future, perhaps due to increasing barriers to entry from technological specialization and higher sunk costs, backward induction becomes possible to turn this strategy upside down. In the last round of the entry game an incumbent would share the industry with an entrant, as there aren't future gains from costly fighting in the present as there are no other firms considering entry to scare off. And in the second last round there's no point fighting as the incumbent plans to share the industry in the next round regardless. This pattern can be traced back round by round to the start, to see an incumbent share industry profits with all entrants from the beginning of the game in any entry game of known finite duration.

Strategic pre-commitment can scare off entrants in both repeated and one-off games of entry, and the process involves an incumbent investing in resources that will only be useful in a price war scenario. For example, in a price war with very low prices equal to marginal cost a firm would be expected to make more sales, and a suitable strategic pre-commitment may be to invest in excess capacity or excess stock. In order for this strategy to work the investment must be both irreversible and visible to the entrant, but if these conditions are met then the game tree transforms into the format below.

Entry deterrence with credible commitment

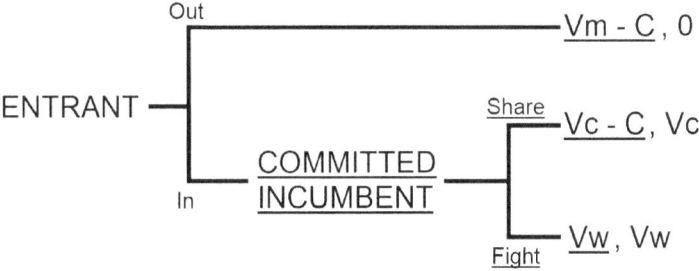

In this game the incumbent is now committed, and has the cost of this commitment (C) subtracted from each possible payoff to represent the cost incurred during the game. For the strategy to work and repel a firm considering entry the fight payoff must exceed the share payoff. This condition should occur as while the cost of the commitment, C, will reduce the value of the share payoff, Vc, as money has been invested that offers nothing in the event the incumbent shares the industry, it will not reduce the value of the fight payoff. The resource costs used for a fight would have been required anyway, and the strategic pre-commitment simply invests in them in advance. This means that in practice Vw - C = Vw for the incumbent, and with a large enough pre-commitment, C, the share payoff should be reduced sufficiently to see it fall below the unchanged fight payoff, and the entrant will know it's now in the best interests of the incumbent to fight, causing them to abandon their entry attempt.

Asymmetric information can also be a big factor in an entry game, as an entrant that doesn't have information about an incumbent firm's costs can easily be deceived. An incumbent may either be a high or low cost producer, and if it's a low cost firm then the entrant may feel they can't possibly compete on prices and avoid entry altogether, but if it's a high cost firm then the entrant may enter feeling sure they can compete and make good profits. The incumbents holding asymmetric information about their true costs can exploit this fact and signal that they're low cost firms, by lowering their prices before a firm enters an industry, even if it costs them in the short-run. A low cost producer would never have high prices as their profit maximization level would occur at a lower price level, and therefore a high price would signal a high cost firm and attract entrants, while a low price would signal a low cost firm and repel those firms considering entry.

Strategic Behaviour

The behaviour for established incumbent firms facing new entrants in their industry can be broken down into four separate strategies:

A Top Dog strategy sees a firm play tough and fight, with an intention to make a rival back off. It could be seen as an active fighting strategy;

The Lean and Hungry Look is based on an understanding that a rival will attack if it senses weakness, and a firm therefore makes sure it doesn't act soft. This could be seen as a passive or indirect fighting strategy;

The Fat Cat Effect is where a firm plays soft and easy, knowing that a rival is likely to follow the example. It could be seen as an active cooperative strategy;

A Puppy Dog Ploy is based on an understanding that a rival firm will fight back if it is attacked, and a firm therefore resolves to avoid acting tough or fighting. This is a passive or indirect cooperative strategy.

A firm could use the lean and hungry look first to repel rivals who may be considering entry into their industry, setting a low or 'limit' price that an entrant couldn't afford to match, and acting as an established firm with low costs and no weaknesses. If an outsider firm is serious about entry then an incumbent could switch to a top dog strategy, engaging in a strategic pre-commitment in excess capacity to prepare for a price war, intending to scare off and intimidate the entrant. If the entrant comes into the industry anyway then an incumbent might adopt a tame puppy dog ploy, to avoid pushing the new firm into a price war that would lower prices and be damaging for the incumbent. And in order to persuade the new entrant into tacit collusion and joint profit maximization, with a share of monopoly profits for all, an incumbent could use the fat cat effect and actively act soft and easy to get along with. This may lead the new entrant into raising its prices to the monopoly level, while maintaining plausible deniability for the incumbent to stay on the right side of the law.

All of these strategies are based on an assumption that the rival firm will react in a certain way. In the case of top dog or the lean and hungry look the assumption is that the rival will follow the opposite strategy, and that a strong top dog play will make the other firm feel weak, while a lean and hungry fighting look will see the other firm avoid a fight. This is similar to Cournot competition where firms compete on output and base their choice of output level upon what they think the other firm will produce, which is assumed to remain fixed. Under Cournot competition if one firm expected the rival firm to produce no output then the first firm would react with a high amount to take the rival firm's market share. And if firm two was expected to produce high output then firm one would react with no output as there would be no point when the other firm will take a large share of the market, and all of the associated industry profits. An output level between the two extremes by a rival firm would be met with an output response between the two by the first firm.

The following diagram shows the Cournot model in visual form, and if firm 1 believes that firm 2 will produce a high level of output then it responds with zero output, shown at the point where the 'firm 1 reaction curve' touches the 'firm 2 output' axis. And if firm 1 believes that firm 2 will produce no output it responds with a high level of output, at the point where the 'firm 1 reaction curve' touches its own 'firm 1 output axis'. These two points are joined by a line to show all possibilities. And this works the opposite way for firm 2, based on the same principle.

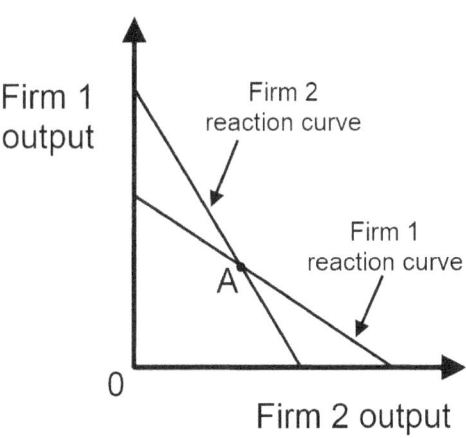

Cournot competition

However, while the firm following a top dog or lean and hungry look strategy may hope that the rival firm will fall for the trick and give up, not entering the industry and not producing any output as they believe that the first firm will produce enough for the whole industry, the strategy may have limited success. In the Cournot model both firms are playing the same game, and the end result is point A in the diagram, where the two reaction curves meet and balance each other out, and where firms share industry output and profits. This is not what the incumbent firm would want.

With fat cat and puppy dog ploy strategies the assumption is that the rival firm will copy a business's strategy and also act cooperative and non-threatening. This idea that a rival firm will copy a firm's own behaviour is like Bertrand competition where firms compete on price. Each firm assumes that the rival will focus on an appealing low price for consumers, as they do, and the firms react to this expected lower price from a rival by reacting with a pre-emptive price cut of their own.

Bertrand price competition is shown in the following diagram, and a firm's price cut reduces their price from the original point of P^*1 for firm one or P^*2 for firm two to a level just below what they expect from the rival firm, which is represented by the dashed 45 degree line. And the price cuts continue downward until they can't be cut any further without affecting the firm's survival, where price equals marginal cost (MC) and there are no profits for anyone.

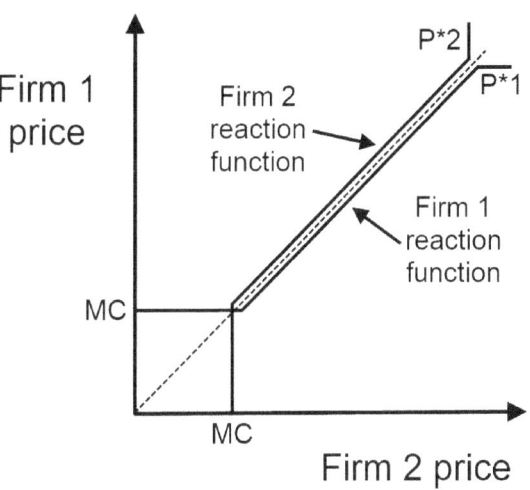

Bertrand competition

Although the top dog, lean and hungry look, puppy dog ploy, and fat cat strategies may have the desired effect and scare off new firms from an industry, these Cournot and Bertrand examples suggest that the strategies may not necessarily work as planned. Depending on the reactions of other firms can put a firm at risk. A business will be best served if it can focus on its own survival and not trying to pre-empt rivals, and this is the subject of the next section.

Firm Survival

The previous sections have shown that a firm has two basic strategies to deal with the current business rivals in their industry and new entrants and survive. It may either fight them (by lowering prices or showing signs of preparation for a price war) in an attempt to keep buyers' sales and money all to themselves, or it can cooperate (by not showing signs of a price war) and share industry profits together. The best outcome for firms would be to fight as the other attempts to cooperate, while the worst payoff would come from being on the other end of this.

It appears that business competition has a great deal in common with the well known hawk-dove game used in evolutionary biology, where animals compete for essential resources (like firms compete for buyer's money to survive). An animal can either act hawk to fight in an attempt to gain all of it (like firms fight with low prices to gain sales), or act dove to avoid the costs of a fight and share the resource instead (as firms act passive to avoid a price war and instead share industry profits). Due to these similarities it may be useful to investigate the hawk-dove game further, and it could offer insights into the best strategy for a player facing a business competition game. The game is shown with its payoffs below.

	Player 2 acts hawk	Player 2 acts dove
Player 1 acts hawk	(V - C) / 2 (V - C) / 2	0 V
Player 1 acts dove	V 0	V / 2 V / 2

If one player acts hawk and the other dove the former's fighting stance scares the passive dove player off, and the fighter gains all of the resource and enjoys its payoff value (V), while the other player gets nothing (0 payoff). If both act dove and cooperate without a fight they share the value of the resource (V / 2). And if both act hawk then the cost of the fight (C) is taken away from the value of the resource before it's shared between players, (V - C) / 2, to represent what a player actually gets from their fight strategy. V will be positive as the resource will have some value, and C will always be above zero as a fight will naturally incur some type of cost. The only question is if V or C will be greater, and whether the value of the resource exceeds the cost of a fight to get it.

If V > C then payoffs would have the following ranking from best to worst:

1st = V
2nd = V / 2

3rd = (V - C) / 2
4th = 0

This will ensure fighting is always best for a Nash equilibrium where both players act as fighters.

	Player 2 acts fighter	Player 2 acts passive
Player 1 acts fighter	<u>3rd</u> 3rd	4th 1st
Player 1 acts passive	1st 4th	2nd 2nd

But it's already known that a V > C condition doesn't hold for firms or they would all engage in price wars and always fight, even beyond the point where price equals marginal cost. The cost of a fight could be the total destruction of a firm, as the fight won't be over until one side wins and this might use up huge amounts of a business's resources, while a sale or several may only add limited money and resources. An accurate portrayal of the hawk-dove game, where players choose between fighting and sharing a needed resource, would see C > V, and the cost of a fight exceeding the value of the prize. This would make the (V - C) payoff in the game negative and below zero, to give the following ranking of payoffs:

1st = V
2nd = V / 2
3rd = 0
4th = (V - C) / 2

And these rankings change the game.

	Player 2 acts fighter	Player 2 acts passive
Player 1 acts fighter	4th 4th	3rd 1st
Player 1 acts passive	1st 3rd	2nd 2nd

This changes the best response to a rival's actions.

	Player 2 acts fighter
Player 1 acts fighter	4th
Player 1 acts passive	3rd

If player two acts as a fighter then player one should now act passive and cooperate, for a higher ranked payoff.

	Player 2 acts passive
Player 1 acts fighter	1st
Player 1 acts passive	2nd

If player two acts passive then player one should act like he will fight, for the superior raked payoff. Overall, player one now doesn't have a dominant strategy, and his best response is to do the opposite of what player two does. This holds for both players as payoffs are identical.

	Player 2 acts fighter	Player 2 acts passive
Player 1 acts fighter	4th 4th	<u>3rd</u> <u>1st</u>
Player 1 acts passive	<u>1st</u> <u>3rd</u>	2nd 2nd

This gives two Nash equilibria, each where one player acts as a fighter (hawk) and one acts passive (dove). At each of these two points neither player can improve their payoff without the other also changing their strategy. Interestingly, this result of divergent strategies for players would also hold if payoffs were asymmetric here, as one player prices the value of resource on offer as greater than the cost of the fight, $V > C$, and the other player values the resource as less than the fight costs, $C > V$. The player who sees the resource as worth fighting for will always choose to fight, and the player who doesn't see it worth fighting for will always do the opposite of the other player, to give the same one fight (hawk) and one passive (dove) outcome as where neither prices the resource above fight costs. All it takes to avoid a fight is for one player to not want it.

It seems strange that the equilibrium here represents unchartered territory, and neither supports firms fighting it out in a price war (both acting hawk), nor firms passively cooperating with tacit collusion for joint profit maximization over the long-run (both acting dove). But this doesn't mean that one player needs to accept they're going to get screwed over and they can't avoid it, and instead the analysis can be seen to suggest that firms should always follow different strategies. In evolutionary biology this would see animals develop their own niche, as one animal would fight for resources in the day as another sleeps passively before they switch roles at night, or one hunts in the trees and a different animal fights at sea. In terms of business competition the game may be giving the same lesson, and the best way for a firm to compete is not to fight it out and beat every other rival firm, nor to compromise and share profits, but to create its own separate niche in the market where it will naturally survive as the fittest firm, as there can be no competition. This appears to be what a firm is trying to do with both Cournot and Bertrand competition, where it aims (and fails) to choose a unique level of output or price that rivals won't match, but there may be better ways to find a niche.

Niche Businesses

The diagrams that follow suggest two different strategies for a business to find its niche in the market, where rival businesses can't compete and greater profit and wealth opportunities are available. Any market may already possess a range of successful firms with differentiated products, and with consumer preferences relatively consistent a firm could find its niche simply by differentiating its products or service to fit in any large gaps between these firms, such as between firms C and D.

Product differentiation niches

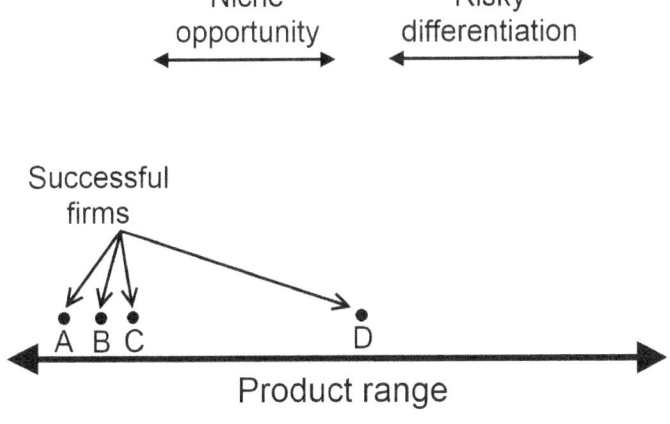

But any point to the right of firm D may be risky and should be avoided, as this is unproven territory where no other firms have had success, and it may not be popular with consumers. And a place between firms A and C is also not possible as there's no space for another firm.

Another way to find a niche is for a firm to take advantage of changing consumer preferences. Other firms may be established and invested in a certain type of product, and unable to quickly react to changing buyer tastes. But a new firm could find a niche if it moved first to satisfy these new preferences.

Changing preferences niche

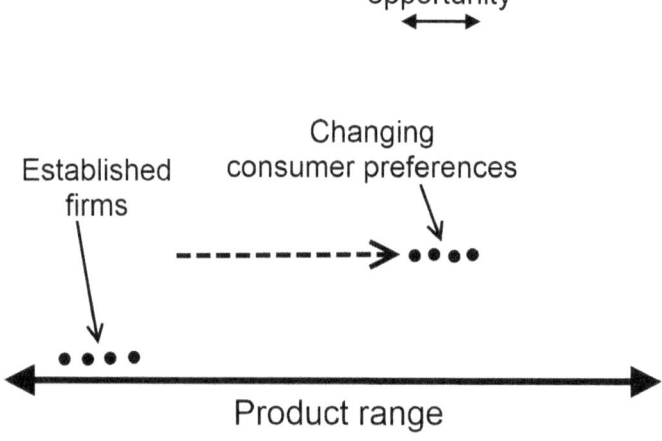

www.ingramcontent.com/pod-product-compliance
Lightning Source LLC
Chambersburg PA
CBHW051732170526
45167CB00002B/907